Skills Worksheet

Concept Review

In the space provided, write the letter of the definition that best matches the term or phrase.

_____ 1. sublimation

_____ 2. precipitation

_____ 3. cloud

_____ 4. dew point

_____ 5. fog

_____ 6. condensation nucleus

_____ 7. latent heat

_____ 8. absolute humidity

_____ 9. relative humidity

_____ 10. coalescence

a. a suspended particle that provides a surface for condensation

b. formation of a large droplet by the combination of small droplets

c. the temperature at which condensation equals evaporation

d. collection of water droplets or ice crystals suspended in the air

e. heat energy that is absorbed or released during a phase change

f. the mass of water vapor contained in a given volume of air

g. the process by which a solid changes directly into a gas

h. a mass of water vapor that condenses near the surface of Earth

i. any form of water that falls to Earth's surface from clouds

j. the ratio of actual water vapor content of the air to the amount of water vapor needed to reach saturation

In the space provided, write the letter of the answer choice that best completes each statement or best answers each question.

_____ 11. What is a low-altitude billowy cloud called?
   a. a stratus cloud
   b. a cumulus cloud
   c. a cirrus cloud
   d. fog

_____ 12. Water vapor changes into a liquid in the process of
   a. evaporation.
   b. supercooling.
   c. condensation.
   d. latent heat.

| Concept Review *continued*

_____ **13.** Precipitation formed in cumulonimbus clouds when convection
currents repeatedly carry raindrops to high levels is
    a. rain.   **c.** snow.
    b. sleet.   **d.** hail.

_____ **14.** The process in which the temperature of an air mass decreases as the
air rises and expands is called
    a. adiabatic cooling.
    b. mixing.
    c. lifting.
    d. advective cooling.

_____ **15.** Fog that results from the nightly cooling of Earth is called
    a. advection fog.
    b. upslope fog.
    c. radiation fog.
    d. steam fog.

_____ **16.** The purpose of cloud seeding is to
    a. predict storms.
    b. induce precipitation.
    c. prevent storms.
    d. prevent condensation.

_____ **17.** A condition in which water is cooled below its freezing point without
going through a change of state is called
    a. sublimation.
    b. condensation.
    c. evaporation.
    d. supercooling.

_____ **18.** In order to find out how intense precipitation will be,
meteorologists use
    a. a rain gauge.
    b. Doppler radar.
    c. cloud seeding.
    d. a psychrometer.

_____ **19.** Large cloud formations associated with storm systems form by
    a. adiabatic cooling.
    b. mixing.
    c. lifting.
    d. advective cooling.

_____ **20.** What forms when the dew point falls below the freezing temperature
of water and water vapor turns directly to ice?
    a. dew   **c.** frozen dew
    b. frost   **d.** sleet

Critical Thinking

Analogies

In the space provided, write the letter of the pair of terms or phrases that best completes the analogy shown. An analogy is a relationship between two pairs of words or phrases written as a:b::c:d. The symbol : is read *is to*, and the symbol :: is read *as*.

_____ **1.** drizzle : rain ::
  a. robin : sparrow
  b. blossom : flower
  c. sand : pebble
  d. rain : snow

_____ **2.** psychrometer : humidity ::
  a. heater : cold
  b. scale : gram
  c. centimeter : meter
  d. thermometer : temperature

_____ **3.** cumulonimbus : thunderstorm ::
  a. seed : plant
  b. ice : water
  c. radiation : fog
  d. water : evaporation

_____ **4.** evaporation : water vapor ::
  a. sublimation : ice
  b. humidity : water
  c. precipitation : cloud
  d. condensation : water

_____ **5.** fog : cloud ::
  a. oak : tree
  b. oak : maple
  c. leaf : tree
  d. acorn : oak

_____ **6.** silver iodide : cloud ::
  a. tree : forest
  b. seed : soil
  c. electricity : light
  d. fan : wind

_____ **7.** stratus : layers ::
  a. cirrus : precipitation
  b. cumulonimbus : storms
  c. cumulus : billows
  d. cirrostratus : halo

_____ **8.** radiation fog : valleys ::
  a. advection fog : coasts
  b. upslope fog : cities
  c. steam fog : oceans
  d. advection fog : rivers

| Critical Thinking *continued*

INTERPRETING OBSERVATIONS

Read the following passage, and answer the questions below.

The city of San Francisco is located on the Pacific coast of northern California. Warm air flows eastward over the Pacific Ocean toward the coast. The California Current, bringing cold water from arctic regions, flows south right along the coast. During the summer, coastal areas of the city experience cool temperatures and much fog. Inland areas of the city have less fog. In September and October, the weather becomes warmer and sunnier and the fog dissipates to a large degree. Sacramento, further inland in the Central Valley, has hot weather and very little fog in the summer. However, in the winter the Central Valley often has fog in the mornings. The table below shows typical summer conditions in the region.

SUMMER CONDITIONS IN NORTHERN CALIFORNIA

	Beach	Downtown San Francisco	Sacramento
Temperature	56°F	62°F	105°F
Dew point	56°F	56°F	56°F

9. What kind of fog occurs in San Francisco in the summer? What causes it?

_____

_____

_____

10. Why do you think the fog lessens in September and October?

_____

_____

_____

11. What kind of fog occurs in the Central Valley in the winter? What causes it?

_____

_____

_____

_____

_____

12. How do the dew point and temperature figures help you understand the summer conditions in the areas described?

_____

_____

_____

_____

_____

_____

AGREE OR DISAGREE

Agree or disagree with the following statements, and support your answers.

13. Meteorologists should not continue to experiment with cloud seeding.

_____

_____

_____

14. Sleet is the form of precipitation most hazardous to people and property.

_____

_____

_____

15. Because of their height, mountaintops tend to have little cloud cover.

_____

_____

_____

_____

16. The burning of fossil fuels could drastically affect the amount of water vapor in the atmosphere.

_____

_____

_____

REFINING CONCEPTS

The statements below challenge you to refine your understanding of concepts covered in the chapter. Think carefully, and answer the questions that follow.

17. Because exercising in high humidity may be uncomfortable, you want to know how humid conditions will be before you play a game of tennis. Would you find it more useful to know the absolute humidity, the relative humidity, or the dew point? Explain why.

_____

_____

_____

_____

18. Imagine that you are a meteorologist and can have only two instruments to help you analyze weather. What two instruments would you choose? Explain your choices.

_____

_____

_____

_____

19. Your home is in a warm dry area known for its clean air. Explain why cloudy days are rare where you live.

_____

_____

_____

_____

_____

20. It is a cold winter day, and you see billowy clouds very high in the sky. What kind of weather would you expect? Explain why.

_____

_____

_____

_____

Skills Worksheet

Directed Reading

Section: Atmospheric Moisture

1. The states in which water exists in the atmosphere are

called _____.

2. The gas phase of water is called _____.

3. The solid phase of water is called _____.

4. The liquid phase of water is called _____.

CHANGING FORMS OF WATER

_____ **5.** When does water change from one phase to another?
  a. when water molecules are held stationary
  b. when evaporation occurs
  c. when heat energy is absorbed or released
  d. when molecules are in a crystalline arrangement

_____ **6.** When ice absorbs energy, the molecules of ice
  a. move more quickly.
  b. become stationary.
  c. become crystals.
  d. slow down.

_____ **7.** What phase does ice change into when it absorbs energy?
  a. gas
  b. liquid
  c. crystals
  d. solid

_____ **8.** When liquid water absorbs energy, it changes to
  a. a gas.
  b. a liquid.
  c. crystals.
  d. a solid.

_____ **9.** What happens to the water molecules when the water absorbs energy?
  a. They move closer together.
  b. They collide more frequently.
  c. They become stationary.
  d. They move more slowly.

Directed Reading *continued*

_____ **10.** The process in which the fastest-moving molecules escape from liquid and form invisible water is called
    a. condensation.
    b. latent heat.
    c. evaporation.
    d. collision.

_____ **11.** The name for heat energy that is absorbed or released during a phase change is
    a. latent heat.
    b. evaporation.
    c. water vapor.
    d. potential energy.

_____ **12.** When liquid water evaporates, the water
    a. releases energy into the atmosphere.
    b. condenses into water vapor.
    c. starts to flow more rapidly.
    d. absorbs energy from the environment.

_____ **13.** What happens to energy absorbed by water during evaporation?
    a. It condenses to form a liquid.
    b. It melts ice.
    c. It is reflected into the atmosphere.
    d. It becomes potential energy between water molecules.

_____ **14.** The name for the process in which water vapor changes back into a liquid is
    a. condensation.
    b. latent heat.
    c. collision.
    d. evaporation.

_____ **15.** During the condensation of water, latent heat
    a. is released into the water.
    b. disappears.
    c. is released into the surrounding air.
    d. is absorbed by the water.

_____ **16.** What happens to latent heat when ice thaws?
    a. It is released.
    b. It is absorbed.
    c. It is recycled.
    d. It is lost.

Directed Reading *continued*

_____ **17.** When water freezes, latent heat
    a. condenses.
    b. is released into the air.
    c. evaporates.
    d. is absorbed.

_____ **18.** Through what process does most water enter the atmosphere?
    a. evaporation
    b. absorption
    c. condensation
    d. release

19. Where on Earth does most evaporation take place?

_____

_____

20. Name four other important sources of water vapor in the atmosphere.

_____

_____

_____

_____

21. How are plants, volcanoes, and burning fuels related to water vapor in the atmosphere?

_____

_____

22. What usually happens to ice before it changes into a gas?

_____

23. What is the name of the process in which a solid changes directly into a gas?

_____

24. Under what conditions might sublimation of snow and ice occur?

_____

_____

25. Water vapor can turn directly into ice without becoming a(n)

_____.

Directed Reading *continued*

HUMIDITY

In the space provided, write the letter of the definition that best matches the term or phrase.

_____ **26.** humidity

_____ **27.** dew point

_____ **28.** absolute humidity

_____ **29.** mixing ratio

a. the temperature at which condensation equals evaporation

b. water vapor in the atmosphere

c. the mass of water vapor contained in a given volume of air

d. the mass of water vapor in a unit of air relative to the mass of the dry air

30. What controls humidity?

_____

31. What determines the rate of evaporation?

_____

32. What happens to the rate of evaporation as the temperature gets higher?

_____

33. What determines the rate of condensation?

_____

34. The part of the total atmospheric pressure that is caused by water vapor is

_____.

35. When there is equilibrium between the rate of evaporation and the rate of

condensation, the air is _____.

36. The measure of the actual amount of water vapor in the air is called the

_____.

37. What equation is used to calculate the absolute humidity?

_____

38. Why do meteorologists prefer to describe humidity by using the mixing ratio of air?

_____

_____

39. What is the mixing ratio of air that has 18 g of water vapor in 1 kg of air?

_____

| Directed Reading *continued*

40. What is a common mixing ratio for air in polar regions?

_____

41. Why is the mixing ratio not affected by changes in temperature or pressure?

_____

_____

42. The ratio of the actual water vapor content of the air to the amount of water vapor needed to reach saturation is called

_____.

43. If a person wanted to know how close the air is to reaching the dew point,

he or she would calculate the _____.

44. At what point does air become saturated at 25°C?

_____

45. How would you express the relative humidity of air that is 25°C and contains 5 g of water?

_____

46. What can make the relative humidity change even if the temperature does not change?

_____

47. What can make the relative humidity increase if the moisture in the air remains the same?

_____

48. What happens to the relative humidity if the temperature increases as the moisture in the air remains constant?

_____

49. What can cause air to cool to its dew point?

_____

_____

50. What is the name of the condensation that forms during the night?

_____

51. What causes dew to form?

_____

_____

| Directed Reading *continued*

52. Under what conditions is dew most likely to form?

_____

53. What is the form of condensation that forms if the dew point falls below the freezing temperature of water?

_____

54. What is the difference between frost and frozen dew?

_____

_____

MEASURING HUMIDITY

In the space provided, write the letter of the definition that best matches the term or phrase.

_____ **55.** dew cell

_____ **56.** electrical conductance

_____ **57.** psychrometer

a. an instrument used to measure relative humidity consisting of two identical thermometers

b. the ability to conduct electricity

c. an instrument used to measure humidity consisting of a heater and two electrodes

58. Why do meteorologists measure humidity?

_____

59. What happens when the lithium chloride in a dew cell absorbs water from the air?

_____

_____

60. What happens as the water evaporates from the LiCl?

_____

_____

61. The temperature at which the LiCl in a dew cell loses its ability to conduct

electricity is the _____.

62. What is the difference between the two thermometers of a psychrometer?

_____

_____

63. What happens to the wet bulb-thermometer when the psychrometer is whirled through the air?

_____

_____

64. How does the temperature of the wet-bulb thermometer differ from that of the dry-bulb thermometer after the psychrometer is whirled through the air?

_____

65. What would you use to calculate the relative humidity from a psychrometer?

_____

In the space provided, write the letter of the definition that best matches the term or phrase.

_____ **66.** hair hygrometer

_____ **67.** radiosonde

_____ **68.** electric hygrometer

a. an instrument that measures humidity at high altitudes

b. an instrument that measures relative humidity by using a bundle of hairs

c. a package that carries instruments into the atmosphere

69. As relative humidity increases, what happens to hair?

_____

_____

70. What is a disadvantage of using a hair hygrometer?

_____

_____

71. How does an electric hygrometer work?

_____

_____

_____

Skills Worksheet

Directed Reading

Section: Clouds and Fog

1. A collection of small water droplets or ice crystals falling slowly through

the air is a(n) _____.

2. The crystals or droplets that make up clouds form when condensation or

sublimation occurs more quickly than the process of _____.

3. A cloud that forms near or on Earth's surface is _____.

CLOUD FORMATION

_____ **4.** What must be available for water vapor to condense and form a cloud?
   a. a solid surface
   b. empty space
   c. high winds
   d. a body of water

_____ **5.** The lowest layer of the atmosphere is the
   a. stratosphere.
   b. ionosphere.
   c. troposphere.
   d. thermosphere.

_____ **6.** What is present in the troposphere that is essential for cloud formation?
   a. a large solid surface
   b. large particles
   c. stationary dust surfaces
   d. tiny suspended particles

_____ **7.** Suspended particles that provide a surface for water vapor to
condense are called
   a. water molecules.
   b. salt molecules.
   c. condensation nuclei.
   d. saturated air.

_____ **8.** What happens when water molecules collect on condensation nuclei?
   a. The rate of condensation decreases.
   b. Water droplets form.
   c. The air temperature reaches the dew point.
   d. The rate of evaporation decreases.

Directed Reading *continued*

_____ **9.** What condition must the air be in for clouds to form?
  a. It must not be saturated with water vapor.
  b. It must have a low relative humidity.
  c. The rate of evaporation must be higher than the rate of condensation.
  d. The rate of condensation must be higher than the rate of evaporation.

_____ **10.** The net condensation that forms clouds may be caused by
  a. the warming of air.
  b. the cooling of air.
  c. rapid evaporation of air.
  d. constant air temperature.

ADIABATIC COOLING

_____ **11.** What happens to molecules in rising air?
  a. They move closer together.
  b. They move farther apart.
  c. They do not move.
  d. They have more collisions.

_____ **12.** What occurs in adiabatic cooling?
  a. Two bodies of moist air mix and change the air temperature.
  b. The temperature of an air mass decreases as the air rises.
  c. Air rises on a mountain and cools.
  d. Air moves over a warm surface and cools.

_____ **13.** What does the adiabatic lapse rate describe?
  a. the temperature of a rising or sinking parcel of air
  b. the amount the temperature of rising or sinking air changes
  c. the amount of clouds in rising or sinking air
  d. the rate at which the temperature of rising or sinking air changes

_____ **14.** What is the adiabatic lapse rate of clear air?
  a. 1°C for every 100 m that air rises
  b. 1°C for every 1000 m that air rises
  c. -1°C for every 100 m that air rises
  d. -0.5°C for every 100 m that air rises

_____ **15.** What is the average adiabatic lapse rate of cloudy air?
  a. more than 1°C per 100 m that air rises
  b. -1°C per 100 m that air rises
  c. between 0.5°C and 0.9°C per 100 m that air rises
  d. between -0.5°C and -0.9°C per 100 m that air rises

Directed Reading *continued*

16. Why does cloudy air have a slower rate of cooling than clear air?

_____

_____

17. What two things happen to the energy from the sun when it reaches Earth's surface?

_____

_____

_____

18. Describe what happens to air near Earth's surface.

_____

_____

_____

19. What is the name of the altitude where net condensation begins to form clouds.

_____

MIXING

20. How does the mixing of two bodies of moist air with different temperatures cause clouds to form?

_____

_____

_____

LIFTING

21. What are the results of air being forced upward?

_____

22. What kind of terrain may force air upward?

_____

23. How do large clouds associated with storm systems form?

_____

_____

ADVECTIVE COOLING

24. What is the name of the process in which the temperature of an air mass decreases as it moves over a cold surface, such as cold ocean or land?

_____

25. What happens when an air mass moves over a surface colder than the air is?

_____

26. What must happen in order for air cooled by adiabatic cooling, mixing, lifting, or advective cooling to form clouds?

_____

_____

CLASSIFICATION OF CLOUDS

27. What two features are used to classify clouds?

_____

_____

28. Name the three basic forms of clouds.

_____

_____

_____

29. What are the three altitude groups of clouds and their heights?

_____

_____

_____

In the space provided, write the letter of the definition that best matches the term or phrase.

_____ **30.** stratus clouds

_____ **31.** altostratus clouds

_____ **32.** cumulus clouds

_____ **33.** cumulonimbus clouds

_____ **34.** cirrus clouds

_____ **35.** cirrostratus clouds

a. feathery clouds composed of ice crystals

b. middle-altitude clouds that usually produce little precipitation

c. high, dark storm clouds

d. clouds that form a high, transparent veil

e. billowy, low-altitude clouds

f. clouds with a flat base forming at very low altitudes

36. Clouds that form where a layer of warm, moist air lies above a layer of cool

air are called _____.

37. What do the prefix *nimbo-* and the suffix *–nimbus* mean?

_____

38. How do nimbostratus clouds differ from other stratus clouds?

_____

39. What does *cumulus* mean?

_____

40. What does the characteristic flat base of cumulus clouds represent?

_____

41. On what two factors does the height of a cumulus cloud depend?

_____

_____

42. In what kind of weather do cumulus clouds grow highest?

_____

43. What are cumulus clouds at middle altitudes called?

_____

44. Name the low clouds that are a combination of two kinds of clouds.

_____

45. What do *cirrus* and *cirro-* mean?

_____

46. At what altitude do cirrus clouds form?

_____

47. Why does light easily pass through cirrus clouds?

_____

48. What kind of clouds often appear before a snowfall or rainfall?

_____

_____

Directed Reading *continued*

FOG

49. Compare and contrast fog and clouds.

_____

_____

In the space provided, write the letter of the description that best matches the term or phrase.

_____ **50.** radiation fog

_____ **51.** advection fog

_____ **52.** upslope fog

_____ **53.** steam fog

a. forms when cool air moves over an inland warm body of water

b. forms due to the loss of heat by radiation when Earth cools at night

c. forms when warm, moist air from above water moves over a cold surface

d. forms when air rises along land slopes

54. Why is radiation fog thickest in valleys and other low places?

_____

_____

55. Why is radiation fog often thick around cities?

_____

_____

56. Where is advection fog common?

_____

_____

Skills Worksheet

Directed Reading

Section: Precipitation

1. Any form of water that falls to Earth's surface from the clouds is

   called _____.

2. Name four major types of moisture that fall from the air to Earth.

   _____

   _____

   _____

   _____

FORMS OF PRECIPITATION

In the space provided, write the letter of the definition that best matches the term or phrase.

_____ **3.** rain

_____ **4.** drizzle

_____ **5.** snow

_____ **6.** sleet

_____ **7.** glaze ice

_____ **8.** ice storm

_____ **9.** hail

a. precipitation consisting of ice particles

b. solid precipitation in the form of lumps of ice

c. a thick layer of ice on a surface

d. clear ice pellets formed when rain falls through a layer of freezing air

e. liquid precipitation

f. rain consisting of drops smaller than 0.5 mm in diameter

g. the condition which produces glaze ice

10. What is the size of normal raindrops?

   _____

11. What is the most common form of solid precipitation?

   _____

12. What are three forms in which snow may fall?

   _____

   _____

   _____

| Directed Reading *continued*

13. How do snowflakes change in size as the temperature goes below 0°C?

_____

14. In what kind of clouds does hail usually form?

_____

15. What process causes hail to form and fall to the ground?

_____

_____

_____

16. Why is hail potentially harmful?

_____

CAUSES OF PRECIPITATION

_____ **17.** The diameter of most cloud droplets is about
    a. 5 millimeters.
    b. 20 micrometers.
    c. 100 micrometers.
    d. 20 millimeters.

_____ **18.** What must happen in order for a cloud droplet to fall as precipitation?
    a. It must freeze.
    b. It must decrease in size.
    c. It must increase in size.
    d. It must warm up.

_____ **19.** What two processes cause cloud droplets to fall to Earth?
    a. coalescence and ultracooling
    b. coagulation and supercooling
    c. coalescence and supercooling
    d. coagulation and superwarming

_____ **20.** What happens in the process of coalescence?
    a. Small droplets slow down as they fall.
    b. Small droplets combine to form larger droplets.
    c. Small droplets break up into smaller droplets.
    d. Large droplets divide into smaller droplets.

_____ **21.** During supercooling, a substance becomes extremely cold and
    a. changes to a solid.
    b. changes to a gas.
    c. changes to a liquid.
    d. does not change its state.

| Directed Reading *continued*

_____ **22.** What is NOT true of freezing nuclei?
    a. They are a form of precipitation.
    b. They are suspended in the air.
    c. They are solid particles.
    d. They are similar to ice in structure.

_____ **23.** Why don't supercooled water droplets freeze?
    a. They are too cold.
    b. They are too large.
    c. There are not enough freezing nuclei available.
    d. There are too many solid particles in the air.

_____ **24.** What does water vapor from supercooled water droplets do?
    a. It condenses on ice crystals that have formed on freezing nuclei.
    b. It evaporates from the freezing nuclei.
    c. Water vapor from the droplets evaporates.
    d. Water vapor makes ice crystals increase in size.

_____ **25.** Which of the following are created by the process of supercooling?
    a. drizzle and rain
    b. sleet and hail
    c. glaze ice and snow
    d. snow and rain

MEASURING PRECIPITATION

26. What is the name of an instrument used to measure rainfall?

_____

27. In one type of _____, a funnel fills one side of a divided

bucket with 0.25 mm of rainwater, and then tips and sets off an electrical
device that records the amount.

28. What instrument measures snow depth?

_____

29. About how much snow does it take to produce 1 cm of water?

_____

30. What does Doppler radar measure?

_____

31. How does Doppler radar work?

_____

_____

Directed Reading *continued*

32. Name three things meteorologists can determine with Doppler radar.

_____

_____

_____

33. How does Doppler radar save lives?

_____

WEATHER MODIFICATION

_____ **34.** The process in which freezing or condensation nuclei are introduced into a cloud to cause rain is called
   a. rain seeding.
   b. cloud seeding.
   c. precipitation growing.
   d. nuclei dropping.

_____ **35.** Which of the following are introduced into a cloud to cause rain because they resemble ice crystals?
   a. snow flakes
   b. hail stones
   c. carbon monoxide pellets
   d. silver iodide crystals

_____ **36.** The substance used in cloud seeding to cool cloud droplets and cause ice crystals to form is
   a. powdered dry ice.
   b. sleet.
   c. water vapor.
   d. snow.

37. What are three ways in which cloud seeding materials are released?

_____

_____

_____

38. Does cloud seeding cause a significant increase in precipitation?

_____

39. What are two ways in which cloud seeding could help people?

_____

_____

Skills Worksheet

Math Skills

Algebraic Rearrangements and Water in the Atmosphere

Algebraic equations contain constants and variables. Constants are specific numbers, such as 2 or 5. Variables are unspecified quantities. They are unknowns. They are represented by letters such as x, y, z, a, b, and c. You will often need to find the value of a variable in a problem. To do so, you can state and then solve the problem as an algebraic equation. Here is an example of an algebraic equation that you would solve to find the value of the variable x.

$$3x = 51$$

In an algebraic equation, the total quantity on one side of the equal sign is equal to the quantity on the other side. If you perform the same operation on each side of the equation, the results will still be equal. To solve an equation, multiply or divide each side of the equation by the same factor, or add or subtract the same amount to or from each side. You can perform any operation on one side of an equation as long as you do the same thing to the other side of the equation. In the example above, you would divide each side of the equation by 3 to find the value of x:

$$3x \div 3 = 51 \div 3$$
$$x = 17$$

SAMPLE PROBLEM:

If 10 cm of snow produces 1 cm of water, how much water does 32 cm of snow produce?

SOLUTION

Step 1: Formulate an equation. Express the information you know as constants. Express the information you want to find out as variables.

$$x = \text{the amount of water}$$
$$10 \text{ cm}/1 \text{ cm} = 32 \text{ cm}/x \text{ or}$$
$$10 = 32 \text{ cm}/x$$

Step 2: Multiply each side of the equation by x.

$$10x = 32 \text{ cm}$$

Step 3: Divide each side of the equation by 10.

$$x = 3.2 \text{ cm}$$

So, 32 cm of snow produces 3.2 cm of water.

Math Skills *continued*

PRACTICE

Using the sample problem as a guide, answer the following questions. Remember to show your work.

1. 10 cm of snow produces approximately 1 cm of water. One winter day in Toronto, 4.3 cm of rain fell. If the precipitation had been snow, how much snow would there have been?

2. The adiabatic lapse rate is -0.6°C per every 100 m that air rises. The air at 0 m is 34°C. At what height would the air reach the dew point of 20°C?

3. Suppose that 1 kg of air can hold up to 18 g of water at 20°C. If 1 kg of air at 20°C holds 7 g of water, what is the relative humidity?

4. The relative humidity is 75%. 1 kg of this air contains 6 g of water vapor. How much water vapor can 1 kg of air hold at the same temperature?

Name _____ Class _____ Date _____

Graphing Skills

Pie Graphs and Cloud Cover

A pie graph shows how the parts of something relate to the whole. Pie graphs are frequently created to show data consisting of percentages. For example, a meteorologist could use a pie graph to show the percentages of different kinds of cloud cover for a particular place over the course of a year.

To make a pie graph, draw a circle to represent the whole or total. Divide the circle into 100 equal sections of 3.6° each. To represent 40% of the whole, shade in 40 consecutive sections. Or you can use a protractor to measure the number of degrees that are represented by a percentage of the circle. For example, 40% would be equal to 40 × 3.6°, or 144°. Mark a section of the circle to represent each percentage. Then use different colors or designs to shade in each portion of the circle.

The table shows the average days per year that each kind of cloud cover was present in Portland, Oregon. The pie graph shows the same information.

CLOUD COVER IN PORTLAND, OREGON

Weather Description	Number of Days per Year
Clear	67
Partly cloudy	71
Cloudy	227

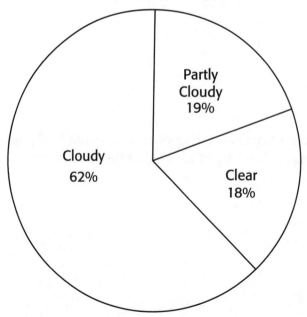

Average Annual Cloud Cover in Portland, Oregon

Name _____ Class _____ Date _____

Graphing Skills *continued*

PRACTICE

Use the table and pie graph to answer the following questions.

1. What is the number of cloudy days in Portland each year? What percentage of the total days in a year is this?

_____

_____

2. What is the number of partly cloudy days in Portland each year? What percentage of days is this?

_____

_____

3. What is the number of clear days in Portland each year? What percentage of days is this?

_____

_____

4. Find the average number of cloudy, partly cloudy, and clear days per year in your town or a nearby city. Create a table and a pie graph to display the information. (You may find this information on the Internet on the National Weather Service Web site or another Web site.)

Cloud Cover in _____

Weather Description	Number of Days per Year	Percentage of Days
Clear		
Partly cloudy		
Cloudy		

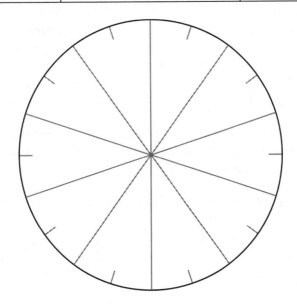

27

Assessment

Section Quiz

Section: Atmospheric Moisture

MATCHING

In the space provided, write the letter of the definition that best matches the term or phrase.

_____ **1.** evaporation

_____ **2.** latent heat

_____ **3.** condensation

_____ **4.** dew

_____ **5.** sublimation

a. the condensation that occurs when air comes into contact with grass and cools

b. the process by which fast-moving molecules escape from a liquid to form water vapor

c. the process by which a solid changes directly into a gas

d. the heat energy absorbed or released by a substance during a phase change

e. the process in which water vapor changes into a liquid

MULTIPLE CHOICE

In the space provided, write the letter of the answer choice that best completes each statement or best answers each question.

_____ **6.** The mass of water vapor contained in a given volume of air is
    a. humidity.
    b. absolute humidity.
    c. relative humidity.
    d. dew point.

_____ **7.** An instrument that measures relative humidity with two identical thermometers is a(n)
    a. dew cell.
    b. hair hygrometer.
    c. radiosonde.
    d. psychrometer.

_____ **8.** From where does most evaporation come?
    a. plants
    b. burning fuels
    c. equatorial oceans
    d. lakes

_____ **9.** The dew point is the temperature at which the rate of condensation
    a. equals the rate of evaporation.
    b. lowers the rate of evaporation.
    c. exceeds the rate of evaporation.
    d. raises the rate of evaporation.

_____ **10.** If the temperature stays the same and the air becomes more moist, the relative humidity will
    a. decrease.
    b. increase.
    c. increase and then decrease.
    d. remain constant.

Section Quiz

Section: Clouds and Fog

MATCHING

In the space provided, write the letter of the description that best matches the term or phrase.

_____ **1.** advection fog

_____ **2.** cirrus clouds

_____ **3.** stratus clouds

_____ **4.** cumulus clouds

_____ **5.** radiation fog

a. results from the nightly cooling of Earth

b. form at low altitudes with a top that resembles cotton balls

c. have the highest altitude of any cloud in the sky

d. forms along coasts when warm, moist air moves across a cold surface

e. cover large areas of sky and often block out the sun

MULTIPLE CHOICE

In the space provided, write the letter of the term or phrase that best completes each statement or best answers each question.

_____ **6.** Where do middle clouds form?
   a. between 1,000 and 2,000 m
   b. between 2,000 and 6,000 m
   c. between 6,000 and 8,000 m
   d. between 8,000 and 10,000 m

_____ **7.** What must be present in order for clouds to form?
   a. adiabatic cooling
   b. a large solid surface
   c. condensation nuclei
   d. a condensation level

_____ **8.** The process by which the temperature of an air mass decreases as the air mass moves over a cold surface is
   a. adiabatic cooling.   **c.** lifting.
   b. mixing.   **d.** advective cooling.

_____ **9.** Upslope fog forms through
   a. adiabatic cooling.   **c.** lifting and cooling.
   b. mixing and warming.   **d.** advective cooling.

_____ **10.** The base of the clouds marks the
   a. adiabatic lapse rate.   **c.** condensation nuclei.
   b. condensation level.   **d.** dew point.

Assessment

Section Quiz

Section: Precipitation

MATCHING

In the space provided, write the letter of the description that best matches the term or phrase.

_____ **1.** supercooling

_____ **2.** rain gauge

_____ **3.** coalescence

_____ **4.** Doppler radar

_____ **5.** cloud seeding

a. introduces condensation nuclei into a cloud

b. measures rainfall amounts

c. combines small cloud droplets into large droplets

d. cools a substance without changing its state

e. measures precipitation intensity

MULTIPLE CHOICE

In the space provided, write the letter of the answer choice that best completes each statement or best answers each question.

_____ **6.** Precipitation in the form of lumps of ice is
   a. sleet.
   b. glaze ice.
   c. drizzle.
   d. hail.

_____ **7.** What have meteorologists concluded about cloud seeding?
   a. It always causes a significant increase in precipitation.
   b. It never produces precipitation.
   c. It sometimes increases and sometimes decreases precipitation.
   d. It has no effect on precipitation.

_____ **8.** The diameter of a normal raindrop is
   a. smaller than 0.5 mm.
   b. between 0.5 and 5 mm.
   c. between 5 mm and 10 mm.
   d. larger than 10 mm.

_____ **9.** What can form if rain falls through a layer of freezing air near the ground?
   a. snow
   b. hail
   c. sleet
   d. big raindrops

_____ **10.** The most common form of solid precipitation is
   a. snow.
   b. sleet.
   c. glaze ice.
   d. hail.

Assessment

Chapter Test A

Chapter: Water in the Atmosphere
MATCHING

In the space provided, write the letter of the definition that best matches the term or phrase.

_____ **1.** absolute humidity

_____ **2.** latent heat

_____ **3.** coalescence

_____ **4.** stratus cloud

_____ **5.** dew point

_____ **6.** supercooling

_____ **7.** cirrus cloud

_____ **8.** sublimation

_____ **9.** relative humidity

_____ **10.** condensation nucleus

a. cooling a substance below its freezing point, condensation point, or sublimation point without a change in state

b. a gray cloud with a flat uniform base

c. heat energy absorbed or released by matter when it changes phase

d. highest altitude, feathery clouds composed of ice crystals

e. ratio of the amount of water vapor in the air to the amount of water vapor needed to reach saturation

f. a solid particle in the atmosphere that provides the surface on which water vapor condenses

g. the actual amount of water vapor contained in a given volume of air

h. process by which ice changes directly into a water vapor

i. formation of a large droplet by the combination of smaller droplets

j. temperature at which the rate of condensation is the same as the rate of evaporation

MULTIPLE CHOICE

In the space provided, write the letter of the answer choice that best completes each statement or best answers each question.

_____ **11.** Fog that is common along coasts, where warm, moist air over the water moves over land is
   a. radiation fog.
   b. advection fog.
   c. upslope fog.
   d. steam fog.

Chapter Test A *continued*

_____ **12.** What might happen if enough energy is absorbed by liquid water?
  a. Condensation will occur.
  b. The water will change to ice.
  c. The water will change to a gas.
  d. The water will never change.

_____ **13.** Compared to the rate for clear air, the adiabatic lapse rate for cloudy air is
  a. slower.
  b. faster.
  c. larger.
  d. the same.

_____ **14.** Some clouds form when a body of moist air combines with another body of moist air with a different temperature is a process called
  a. adiabatic cooling.
  b. mixing.
  c. lifting.
  d. advective cooling.

_____ **15.** An instrument that measures precipitation by bouncing radio waves off rain or snow is
  a. a hair hygrometer.
  b. radiosonde.
  c. Doppler radar.
  d. a rain gauge.

_____ **16.** A cloud whose name has the prefix *nimbo*- or the suffix *–nimbus* is
  a. high.
  b. layered.
  c. precipitation-free.
  d. rain-producing.

_____ **17.** In cloud seeding, silver-iodide crystals are used as
  a. heating elements.
  b. freezing nuclei.
  c. dew cells.
  d. dry ice.

_____ **18.** The mass of water vapor in a unit of air relative to the mass of the dry air is called the
  a. absolute humidity.
  b. relative humidity.
  c. adiabatic lapse rate.
  d. mixing ratio.

_____ **19.** When the air temperature decreases, the rate of evaporation
  a. decreases.
  b. increases.
  c. may increase or decrease.
  d. remains constant.

_____ **20.** A drop of liquid precipitation that is 2 mm in diameter is
  a. drizzle.
  b. rain.
  c. sleet.
  d. hail.

Assessment

Chapter Test B

Chapter: Water in the Atmosphere

MATCHING

In the space provided, write the letter of the definition that best matches the term or phrase.

_____ **1.** advective cooling

_____ **2.** sublimation

_____ **3.** supercooling

_____ **4.** adiabatic cooling

_____ **5.** coalescence

a. cooling a substance below its freezing point, condensation point, or sublimation point without changing its state

b. decrease in temperature of an air mass as the air mass moves over a cold surface

c. process in which small droplets join to form a large droplet

d. decrease in temperature of an air mass as the air rises and expands

e. changing of a solid directly into a gas

MULTIPLE CHOICE

In the space provided, write the letter of the answer choice that best completes each statement or best answers each question.

_____ **6.** When ice melts, latent heat
    a. is released.
    b. is absorbed.
    c. evaporates.
    d. is sublimated.

_____ **7.** Clouds that often bring thunderstorms are
    a. nimbostratus.
    b. stratocumulus.
    c. altocumulus.
    d. cumulonimbus.

_____ **8.** Precipitation that occurs when rain falls through a layer of freezing air close to the ground is known as
    a. hail.
    b. drizzle.
    c. sleet.
    d. snow.

_____ **9.** Condensation nuclei are
   a. ice and dust particles.
   b. large solid surfaces.
   c. bodies of moist air.
   d. icy clouds.

_____ **10.** Which of the following accurately describes the dew point?
   a. The rate of evaporation exceeds the rate of condensation.
   b. The rate of condensation is higher than the rate of evaporation.
   c. The rate of evaporation and the rate of condensation are equal.
   d. The vapor pressure is high and the condensation rate is low.

_____ **11.** The average amount of water produced by 50 cm of snow is
   a. 1 cm.
   b. 5 cm.
   c. 10 cm.
   d. 50 cm.

_____ **12.** Based on results from cloud seeding so far, meteorologists will most likely
   a. advise against cloud seeding because it is risky.
   b. expand seeding efforts because it is successful.
   c. stop seeding because it does not work.
   d. continue experimenting because the results are mixed.

_____ **13.** Technology that can save lives by warning people of an approaching storm is
   a. radiosonde.
   b. a Psychrometer.
   c. a rain gauge.
   d. Doppler radar.

_____ **14.** Where would the air contain the most moisture?
   a. over Hawaii
   b. over Arizona
   c. over the Arctic Circle
   d. over the Rocky Mountains

_____ **15.** The mass of water vapor in a unit of air relative to the mass of the dry air is
   a. humidity.
   b. the relative humidity.
   c. the absolute humidity.
   d. the mixing ratio.

Chapter Test B *continued*

COMPLETION

Complete each statement by writing the correct term or phrase in the space provided.

16. The altitude at which net condensation causes clouds to form is called

the _____.

17. The process that causes cloud formation when a moving mass of air meets

a mountain range is _____.

18. The clouds with the highest altitude of any clouds in the sky

are_____.

19. The process of cloud formation that occurs when one body of moist air combines with another body of moist air with a different temperature is

called _____.

20. Clouds that begin to form at very low altitudes, covering large areas of sky

and often blocking out the sun, are _____.

SHORT ANSWER

Read each question or statement, and write your answer in the space provided.

21. What is the difference between absolute humidity and relative humidity?

_____

_____

_____

_____

_____

22. Why does dew often form at night?

_____

_____

_____

_____

_____

_____

23. What does a hair hygrometer measure? On what principle is a hair hygrometer based?

_____

_____

_____

_____

_____

CRITICAL THINKING

24. You are camping in a valley on a calm, clear night. When you awake you are surrounded by a thick fog. What type of fog is it, and how did it form?

_____

_____

_____

_____

25. You see lumps of ice the size of golf balls falling from the sky. What form of precipitation are you seeing? Explain how it formed.

_____

_____

_____

_____

_____

Skills Practice Lab

Relative Humidity

Earth's atmosphere acts as a reservoir for water that evaporates from Earth's surface. However, the amount of water vapor that the atmosphere can hold depends on the relative rates of condensation and evaporation. When the rates of condensation and evaporation are equal, the air is said to be "saturated." However, when the rate of condensation exceeds the rate of evaporation, water droplets begin to form in the air or on nearby surfaces. The point at which the condensation rate equals the evaporation rate is called the *dew point* and depends on the temperature of the air and on the atmospheric pressure.

Relative humidity is the ratio of the amount of water vapor in the air to the amount of water vapor that is needed for the air to become saturated. This ratio is most commonly expressed as a percentage. When the air is saturated, the air is said to have a relative humidity of 100%. In this lab, you will use wet-bulb and dry-bulb thermometer readings to determine the relative humidity of the air in your classroom.

OBJECTIVES

Measure humidity in the classroom.

Calculate relative humidity.

MATERIALS

- cloth, cotton, at least 8 cm × 8 cm
- container, plastic
- piece of paper
- ring stand with ring

- rubber band
- string
- thermometer, Celsius (2)
- water

SAFETY

PROCEDURE

1. Hang two thermometers from a ring stand.

2. Using a rubber band, fasten a piece of cotton cloth around the bulb of one thermometer. Adjust the length of the string so that only the cloth, not the thermometer bulb, is immersed in the water. By using this setup, you can measure both the air temperature and the cooling effect of evaporation.

3. Predict whether the two thermometers will have the same reading or which thermometer will have the lower reading.

_____

_____

37

4. Using a piece of paper, fan both thermometers rapidly until the reading on the wet-bulb thermometer stops changing. Read the temperature on each thermometer.

   a. What is the temperature on the dry-bulb thermometer?

_____

_____

   b. What is the temperature on the wet-bulb thermometer?

_____

_____

   c. What is the difference in the two temperature readings?

_____

_____

5. Use the table entitled "Relative Humidity" in the Reference Tables section of the Appendix in your textbook to find the relative humidity based on your temperature readings in **Step 4**. Look at the left-hand column labeled "Dry-Bulb Temperature." First, find the temperature you recorded in **Step 4a**. Follow along to the right in the table until you come to the number that is directly below the column entitled "Difference in Temperature" (top row of the table) and that you recorded in **Step 4c**. This number, expressed as a percentage, is the relative humidity. What is the relative humidity of the air in your classroom?

_____

_____

ANALYSIS AND CONCLUSION

1. Drawing Conclusions On the basis of the relative humidity you calculated, is the air in your classroom close to or far from the dew point? Explain your answer.

_____

_____

2. Applying Conclusions If you wet the back of your hand, would the water evaporate and cool your skin?

_____

_____

_____

Relative Humidity *continued*

EXTENSION

1. **Making Inferences** Suppose that you exercise in a room in which the relative humidity is 100%.

   a. Would the moisture on your skin from perspiration evaporate easily?

   _____

   _____

   b. Would you be able to cool off readily? Explain your answer.

   _____

   _____

   _____

   _____

2. **Applying Ideas** Suppose that you have just stepped out of a swimming pool. The relative humidity is low, about 30%. Would you feel warm or cool? Explain your answer.

   _____

   _____

   _____

   _____

Quick Lab

Dew Point

MATERIALS

- water (room-temperature)
- glass container
- ice cubes (one or two)

SAFETY

PROCEDURE

1. Pour **room-temperature water** into a **glass container**, such as a drinking glass, until the water level is near the top of the cup.

2. Observe the outside of the glass container, and record your observations.

_____

_____

3. Add **one or two ice cubes** to the container of water.

4. Watch the outside of the container for five minutes for any changes.

ANALYSIS

1. What happened to the outside of the container?

_____

_____

2. What is the liquid on the container?

_____

_____

3. Where did the liquid come from? Explain your answer.

_____

_____

_____

Quick Lab

Cloud Formation

MATERIALS

- bottle opener
- glass jar
- 1 mL of hot water
- ice cube

PROCEDURE

1. Use a **bottle opener** to puncture 1 or 2 holes into the lid of a **glass jar.**
2. Pour **1 mL of hot water** into the jar.
3. Place an **ice cube** over the holes in the lid of the jar. Make sure the holes are completely covered.
4. Observe the changes that occur within the jar.

ANALYSIS

1. Draw a diagram of the jar. Label the areas of the diagram where evaporation and condensation take place. Also, label areas where latent heat is released and absorbed.

2. Explain why latent heat was released and absorbed in the areas you labeled on the diagram.

_____

_____

_____

Inquiry Lab

How Big Is a Raindrop?

Have you ever been caught in the rain? If you have, then you know that some raindrops are smaller than others. When it's drizzling, most of the drops seem to be very small. When there is a downpour, most of the drops seem to be larger. So, how big is a raindrop? Are all the drops that fall during a particular episode of rain all the same size? In this lab, you will discover some answers by doing an experiment with real raindrops.

OBJECTIVES

Form a hypothesis about the size of raindrops.

Investigate the size of raindrops by performing an experiment.

Observe raindrops.

Collect data about real raindrops as they fall.

Draw conclusions about the size of raindrops.

MATERIALS

- beaker or cup
- dropper
- flour (approximately1 cup)
- meterstick or ruler
- pan, metal or plastic
- paper plate or laminated placemat
- sifter or strainer, flour
- smock or old shirt, one for each group member
- water

SAFETY

ASK A QUESTION

1. How big is a raindrop?

FORM A HYPOTHESIS

2. Form a hypothesis that answers your question. Explain your reasoning.

_____

_____

_____

How Big is a Raindrop? *continued*

TEST THE HYPOTHESIS

3. Wait for a rainy day to perform this experiment.

4. Wear a smock or old shirt to protect your clothing from the flour.

5. Do all the following steps with the pan on the placemat or paper plate. Sift the flour into the pan evenly. When you are finished, the flour should be about one inch deep in the pan. The surface of the flour should be nearly level. **CAUTION:** Spilled flour may cause you to slip and fall. Wipe up any spilled flour immediately.

6. Before you do the experiment with real raindrops, do a test run with artificial raindrops. Fill the dropper with water. Squeeze the dropper slightly, so that a drop of water forms, *but does not fall.* Measure the drop as best you can with the meterstick or ruler. Record your measurement.

_____

_____

_____

7. After you record your measurement, hold the dropper as high as you can above the pan. Gently squeeze the dropper until the drop falls. What happens when the drop hits the flour in the pan?

_____

_____

_____

_____

8. Observe the dough pellet. How does the dough pellet compare in size to the drop of water?

_____

_____

_____

_____

9. Repeat Steps 6, 7 and 8 two more times. What did you observe about the results when you repeated the experiment?

_____

_____

10. Remove the dough pellet from the pan and set it aside. Place the flour pan out in the rain for four or five seconds, then bring it inside again. What happened to the flour in the pan when it was exposed to the raindrops?

_____

_____

_____

_____

11. Measure each of the pellets that formed and record the measurements in the space below.

_____

_____

_____

_____

ANALYZE THE RESULTS

1. Analyzing Results How did the dough pellets produced by the raindrops compare in size to each other?

_____

_____

_____

_____

2. Explaining Events What might explain the results you obtained for the sizes of the dough pellets?

_____

_____

_____

_____

_____

_____

How Big is a Raindrop? *continued*

DRAW CONCLUSIONS

3. Drawing Conclusions What was the purpose of doing the practice experiment with the dropper before doing the experiment with real raindrops?

_____

_____

_____

_____

4. Making Predictions If you were to repeat the experiment during another rain shower, would you expect the raindrop-size results to be the same? Explain.

_____

_____

_____

_____

EXTENSION

1. Long-term Investigation and Communication Repeat the experiment on several other rainy days. Keep track of the results of the sizes of the raindrops. Make a poster describing what you did and illustrate your results by creating a data chart, and making drawings of the size of raindrops.

_____

_____

_____

_____

_____

_____

_____

_____

Name _____ Class _____ Date _____

What Is the Shape of a Raindrop?

What shape do you think of when you picture a raindrop? Most people picture a teardrop shape, wide on the bottom and pointy on the top. Is that the actual shape of a raindrop? In this experiment, you will create model raindrops and observe their shapes. You will find out the true shape of a raindrop.

OBJECTIVES

Using Scientific Methods **Model** the way raindrops behave in a cloud.

Investigate the shape of raindrops by performing an experiment.

Using Scientific Methods **Observe** the shape, or shapes, of raindrops.

MATERIALS

- beaker or cup
- cheesecloth or paper towel
- clamp
- dropper
- plastic flexible pipe

- ring stand
- rubber band
- vacuum cleaner or hair dryer
- water
- wide clear tape or duct tape

SAFETY

PROCEDURE

1. Attach the plastic pipe to the exhaust port of the vacuum cleaner, or the hair dryer set on "cold." Attach the open end of the pipe to the ring stand, using the clamp. Place a layer of cheesecloth or paper towel over the opening of the plastic pipe. Secure the cheesecloth or paper towel with the rubber band. Tape the vacuum tube to the base of the ring stand or countertop to keep the force of the air from flipping the ring stand.

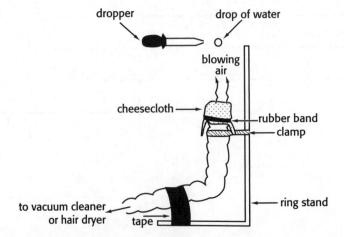

What is the Shape of a Raindrop? *continued*

2. Turn on the vacuum cleaner or hair dryer.

3. CAUTION: Never place drops of water directly over an electrical appliance. There is a danger of electrocution.

Fill the dropper with water. Squeeze one drop of water into the air stream. The drop should be suspended in the air stream, just as a drop of water in a cloud would be suspended by upward air currents. If the air stream is too strong to suspend a drop, place another layer of cheesecloth or paper toweling over the opening of the plastic pipe.

CAUTION: Use only water droplets in the wind stream as directed —do not use any other materials. Wipe up any spilled water immediately.

4. Observe the shape of the "raindrop." Write your observations below.

_____

_____

_____

_____

_____

_____

5. Draw a small illustration to show the shape of the raindrop.

What is the Shape of a Raindrop? *continued*

6. Repeat the experiment two more times. Report your observations in the space below.

_____

_____

_____

_____

ANALYSIS AND CONCLUSION

1. **Explaining Events** Describe what happens when you place a drop of water into the air stream.

_____

_____

_____

_____

2. **Analyzing Results** Did the results of the experiment match the shape that you thought it would be? Explain.

_____

_____

_____

3. **Evaluating Models** How do you think this model compares to the natural processes that determine the shape of a raindrop? Explain.

_____

_____

_____

_____

What is the Shape of a Raindrop? *continued*

EXTENSION

1. **Making Models** Repeat the experiment using different-sized drops of water, and/or different air speeds. Report the results to your class.

_____

_____

_____

_____

2. **Research and Communication** Many scientists are interested in studying the shape of raindrops. NASA has even done experiments about the shape of raindrops in microgravity on the International Space Station. Do research about the shape of raindrops on the internet or in the library. Report your results to the class.

_____

_____

_____

_____

Name _____ Class _____ Date _____

Internet Activity

Latent Heat and Thunderstorms

Use an Internet search engine to find information about the role of latent heat in thunderstorms and other severe weather events.

1. What search engine did you use?_____

_____

2. What keywords did you use to search for information about this topic?

_____

3. List the URLs for two Web sites about this topic. _____

_____

4. What is the source for the information on these Web sites? _____

_____

5. List any universities or professional organizations with which these Web

sites are affiliated. _____

_____

Use the Web sites you listed in question 3 to answer the questions that follow.

6. Explain the part that latent heat plays in creating thunderstorms.

_____

_____

7. Explain the part that latent heat plays in creating tornadoes.

_____

_____

GOING FURTHER

8. **Applying Ideas** Create a diagram that illustrates the patterns of air currents and moisture in a thunderstorm, severe thunderstorm, or tornado.

Activity

Internet Activity

Climate and Precipitation

Use an Internet search engine to find information about the climate of a region in the United States. Find the annual rates of two forms of precipitation in the region.

1. What search engine did you use?_____

_____

2. What keywords did you use to search for this Web site?_____

_____

3. List the URLs for two Web sites about this topic. _____

_____

4. What is the source for the information on these Web sites? _____

_____

5. List any universities or professional organizations with which these Web

sites are affiliated. _____

_____

Use the Web sites you listed in question 3 to answer the questions that follow.

6. Name the region you researched. _____

7. Name one form of precipitation you researched. What is the annual amount of

this form of precipitation in the region?_____

_____

8. Name the second form of precipitation. What is the annual amount of this

form of precipitation in the region?_____

_____

GOING FURTHER

9. Analyzing Relationships Identify the most distinctive features of the region's precipitation patterns, such as heavy rain, heavy snow, or little rain. Describe factors that are responsible for these patterns, such as large bodies of water, elevation levels, or air temperature, and explain how they are responsible.

Answer Key

Concept Review

1. G	**11.** B
2. I	**12.** C
3. D	**13.** D
4. C	**14.** A
5. H	**15.** C
6. A	**16.** B
7. E	**17.** D
8. F	**18.** B
9. J	**19.** C
10. B	**20.** B

Critical Thinking

1. C

2. D

3. A

4. D

5. A

6. B

7. C

8. A

9. Advection fog; warm moist air from the Pacific cools as it moves across the California Current. Advection fog is common along coasts.

10. Answers may vary. Sample answer: It probably lessens because the California Current and the land on the coast have warmed up somewhat by September.

11. Radiation fog, which results from the loss of heat by radiation and is thickest in valleys.

12. Answers may vary. Sample answer: At the beach the temperature is at the dew point, so condensation can occur with just a little cooling. In downtown San Francisco more cooling must take place to reach the dew point. Further inland, the temperature is well above the dew point and so is less likely to reach the dew point.

13. Answers may vary. Sample answer: Disagree. Cloud seeding results are inconclusive and further experimentation is needed.

14. Answers may vary. Sample answer: Disagree. Hail is much more damaging than sleet.

15. Answers may vary. Sample answer: Disagree. The process of lifting, which results when air meets the sloping terrain of mountains, causes air to cool, expand, and form clouds that can cover mountaintops.

16. Answers may vary. Sample answer: Disagree. Only small amounts of moisture are added to the atmosphere from fuels.

17. Answers may vary. Sample answer: the relative humidity because it tells how nearly saturated the air is at a given temperature. The absolute humidity would only tell the amount of vapor, not how saturated the air is. The dew point by itself wouldn't tell you the humidity.

18. Answers may vary. Sample answer: A psychrometer, because it can measure temperature and humidity, and Doppler radar, because it predicts storms in detail.

19. Answers may vary. Sample answer: Clouds are formed by condensation of water vapor. Condensation rates would tend to be lower in warm, dry areas. Clean air would have fewer of the particles needed for water vapor to condense and form clouds.

20. You could expect snow because the cloud is a cirrocumulus, which commonly appear before snow or rain.

Directed Reading

SECTION: ATMOSPHERIC MOISTURE

1. phases

2. water vapor

3. ice

4. water

5. C

6. A

7. B

8. A

9. B

10. C

11. A
12. D
13. D
14. A
15. C.
16. B.
17. B
18. A
19. in oceans of the equatorial regions
20. evaporation from lakes, ponds, streams, and soil
21. They release small amounts of water vapor into the atmosphere.
22. It changes into a liquid.
23. sublimation
24. when the air is dry and the temperature is below freezing
25. liquid
26. B
27. A
28. C
29. D
30. rates of condensation and evaporation
31. air temperature
32. It gets higher.
33. vapor pressure
34. vapor pressure
35. saturated
36. absolute humidity
37. absolute humidity = mass of water vapor (grams) / volume of air (cubic meters)
38. because, as air moves, its volume changes as a result of temperature and pressure changes.
39. 18 g/kg
40. less than 1 g/kg
41. because the measurement uses only units of mass, not units of volume
42. relative humidity
43. relative humidity
44. when it contains 20 g of water vapor per 1 kg of air
45. 25%
46. moisture entering the air
47. a decrease in the temperature
48. The relative humidity will decrease.
49. conduction when the air is in contact with a cold surface
50. dew
51. Objects near the ground lose heat during the night, often dropping to the dew point of the surrounding air. Air comes into contact with the surfaces and water vapor condenses out of it.
52. Cool clear nights with little wind
53. frost
54. Frost forms when water vapor turns directly to ice; frozen dew forms when dew freezes as clear beads of ice.
55. C
56. B
57. A
58. It helps them predict weather conditions.
59. Its electrical conductance increases.
60. The LiCl loses its ability to conduct electricity.
61. dew point
62. The bulb of one is covered with a damp wick and the bulb of the other remains dry.
63. The water in the wick evaporates, and so heat is withdrawn from that thermometer.
64. That of the wet-bulb thermometer is lower.
65. the difference between the readings of the two thermometers
66. B
67. C
68. A
69. It gets longer.
70. It is less accurate than psychrometers and dew cells.
71. An electric current is passed through a moisture-attracting substance. The amount of moisture changes the electrical conductivity of the substance and can be measured and expressed as relative humidity of the surrounding air.

SECTION: CLOUDS AND FOG

1. cloud
2. evaporation
3. fog
4. A
5. C
6. D
7. C
8. B
9. D
10. B
11. B
12. B

13. D
14. C
15. D
16. because of the release of latent heat as the water condenses
17. Earth's surface absorbs it and then reradiates it as heat.
18. It absorbs heat. It rises, expands, and then cools.
19. condensation level
20. The combination causes the temperature of the air to change. The combined air may be cooled to below its dew point, which results in clouds.
21. the cooling of the air and cloud formation
22. sloping terrain, such as a mountain range
23. A mass of cold, dense air enters an area and pushes a less dense mass of warmer air upward.
24. advective cooling
25. The cold surface absorbs heat from the air and the air cools.
26. The air must be cooled to below its dew point.
27. shape and altitude
28. stratus, cumulus, cirrus
29. low clouds (0 to 2,000 m); middle clouds (2,000 to 6,000 m); high clouds (above 6,000 m)
30. F
31. B
32. E
33. C
34. A
35. D
36. stratus
37. "rain"
38. They can cause heavy precipitation.
39. "piled" or "heaped"
40. the condensation level
41. on the stability of the troposphere and on the amount of moisture in the air
42. hot, humid days
43. altocumulus clouds
44. stratocumulus clouds
45. "curly"
46. at altitudes above 6,000 m
47. because they are thin
48. cirrocumulus
49. Fog and clouds are both the result of the condensation of water vapor in the air, but fog is much nearer Earth's surface than clouds and forms differently.
50. B
51. C
52. D
53. A
54. because dense, cold air sinks to low elevations
55. Smoke and dust particles act as condensation nuclei.
56. along coasts

SECTION: PRECIPITATION

1. precipitation
2. rain, snow, sleet, and hail
3. E
4. F
5. A
6. D
7. C
8. G
9. B
10. 0.5 to 5 mm in diameter
11. snow
12. as small pellets, as individual crystals, or as crystals combining to form snowflakes
13. They get smaller.
14. cumulonimbus
15. Convection currents within clouds carry raindrops to high levels, where they freeze. They are carried upward again, accumulating additional layers of ice. They fall when they are too heavy for the convection currents to carry.
16. It can damage crops and property.
17. B
18. C
19. C
20. B
21. D
22. A
23. C
24. A
25. D
26. rain gauge
27. rain gauge.
28. measuring stick
29. 10 cm
30. the intensity of precipitation

31. It bounces radio waves off rain or snow and times how long the wave takes to return.
32. location, direction of movement, and intensity of precipitation
33. by warning people of approaching storms
34. B
35. D
36. A
37. burners on the ground, flares dropped from aircraft, or dropping from aircraft
38. Sometimes cloud seeding produces more rain and at other times it does not increase precipitation.
39. It could ease drought and it could help control severe storms.

Math Skills

1. x = the amount of snow
   $1/10 = 4.3/x$ (Multiply both sides by x.)
   $1/10x = 4.3$ (Divide both sides by 1/10.)
   $x = 43$ cm
2. x = the height at the dew point of 20°
   $34° - 20° = 14°$
   $-0.6°/100 = -14°/x$ (Multiply both sides by x.)
   $-0.6x/100 = -14$ (Multiply both sides by 100.)
   $-0.6x = -1400$ (Divide both sides by -0.6.)
   $x = 2,333$ m
3. x = relative humidity
   $18x = 7$ (Divide both sides by 18.)
   $x = 38.9\%$
4. x = total amount of water vapor air can hold
   $6 = 75/100x$ (Multiply both sides by 100.)
   $600 = 75x$ (Divide both sides by 75.)
   $x = 8g$

Graphing Skills

1. 227 days; 62%
2. 71 days; 19%
3. 67 days; 18%
4. Answers may vary. Tables and pie graphs should look similar to the example.

Section Quizzes

SECTION: ATMOSPHERIC MOISTURE
1. B
2. D
3. E
4. A
5. C
6. B
7. D
8. C
9. A
10. B

SECTION: CLOUDS AND FOG
1. D
2. C
3. E
4. B
5. A
6. B
7. C
8. D
9. C
10. B

SECTION: PRECIPITATION
1. D
2. B
3. C
4. E
5. A
6. D
7. C
8. B
9. C
10. A

Chapter Test A

1. G
2. C
3. I
4. B
5. J
6. A
7. D
8. H
9. E
10. F
11. B
12. C
13. A
14. B

15. C
16. D
17. B
18. D
19. A
20. B

have been carried up over and over again, adding layers until it became to heavy to be carried by the currents

Chapter Test B

1. B
2. E
3. A
4. D
5. C
6. B
7. D
8. C
9. A
10. C
11. B
12. D
13. D
14. A
15. D
16. condensation level
17. lifting
18. cirrus clouds
19. mixing
20. stratus clouds
21. Absolute humidity is the mass of water vapor contained in a given volume of air, while relative humidity is a ratio of the water vapor content of air to the amount of water vapor needed to reach saturation at a given temperature.
22. Because grass and other objects near the ground lose heat at night, cooling to the dew point. The air cools to the dew point when it comes into contact with the ground. Then, water vapor condenses on these surfaces.
23. Humidity; it works on the principle that hair gets longer as relative humidity increases.
24. radiation fog, which forms when cold air sinks to low elevations and is chilled by the ground to below its dew point, then its water vapor condenses into droplets.
25. hail, which forms when convection currents in cumulonimbus clouds carry raindrops to high levels where they freeze; golf-ball-sized hail must

Relative Humidity

Teacher's Notes

TIME REQUIRED One 45-minute class

Randa Flinn
Northeast
High School
Fort Lauderdale, Florida

LAB RATINGS   Easy ◄—— 1   2   3   4 ——► Hard

Teacher Prep–1
Student Set-Up–2
Concept Level–2
Clean Up–1

SKILLS ACQUIRED

Predicting
Experimenting
Collecting Data
Interpreting Results

THE SCIENTIFIC METHOD

In this lab, students will

• Make Observations

• Analyze Results

• Communicate Results

MATERIALS

The materials listed are enough for groups of 2 to 4 students. A rubber band or piece of string will keep the cloth securely fastened around the wet-bulb thermometer. If ring stands are unavailable, thermometers can also be securely mounted on a piece of stiff poster board by using tape. This also makes transporting and using the psychometer later in an outside weather shelter easier.

TIPS AND TRICKS

The wet-bulb is chilled because it releases heat to the evaporating water. The drier the air is, the faster the water will evaporate, and the more the bulb will be chilled.

Students will need to take readings from both bulbs to obtain the relative humidity from the Relative Humidity Table in the Appendix.

Skills Practice Lab

Relative Humidity

Earth's atmosphere acts as a reservoir for water that evaporates from Earth's surface. However, the amount of water vapor that the atmosphere can hold depends on the relative rates of condensation and evaporation. When the rates of condensation and evaporation are equal, the air is said to be "saturated." However, when the rate of condensation exceeds the rate of evaporation, water droplets begin to form in the air or on nearby surfaces. The point at which the condensation rate equals the evaporation rate is called the *dew point* and depends on the temperature of the air and on the atmospheric pressure.

Relative humidity is the ratio of the amount of water vapor in the air to the amount of water vapor that is needed for the air to become saturated. This ratio is most commonly expressed as a percentage. When the air is saturated, the air is said to have a relative humidity of 100%. In this lab, you will use wet-bulb and dry-bulb thermometer readings to determine the relative humidity of the air in your classroom.

OBJECTIVES

Measure humidity in the classroom.

Calculate relative humidity.

MATERIALS

- cloth, cotton, at least 8 cm × 8 cm
- container, plastic
- piece of paper
- ring stand with ring
- rubber band
- string
- thermometer, Celsius (2)
- water

SAFETY

PROCEDURE

1. Hang two thermometers from a ring stand.

2. Using a rubber band, fasten a piece of cotton cloth around the bulb of one thermometer. Adjust the length of the string so that only the cloth, not the thermometer bulb, is immersed in the water. By using this setup, you can measure both the air temperature and the cooling effect of evaporation.

3. Predict whether the two thermometers will have the same reading or which thermometer will have the lower reading.

   Answers may vary.

   _____

   _____

4. Using a piece of paper, fan both thermometers rapidly until the reading on the wet-bulb thermometer stops changing. Read the temperature on each thermometer.

   a. What is the temperature on the dry-bulb thermometer?

   Answers may vary. _____

   _____

   b. What is the temperature on the wet-bulb thermometer?

   Answers may vary. _____

   _____

   c. What is the difference in the two temperature readings?

   Answers may vary. _____

   _____

5. Use the table entitled "Relative Humidity" in the Reference Tables section of the Appendix in your textbook to find the relative humidity based on your temperature readings in **Step 4**. Look at the left-hand column labeled "Dry-Bulb Temperature." First, find the temperature you recorded in **Step 4a**. Follow along to the right in the table until you come to the number that is directly below the column entitled "Difference in Temperature" (top row of the table) and that you recorded in **Step 4c**. This number, expressed as a percentage, is the relative humidity. What is the relative humidity of the air in your classroom?

_____

ANALYSIS AND CONCLUSION

1. Drawing Conclusions On the basis of the relative humidity you calculated, is the air in your classroom close to or far from the dew point? Explain your answer.

   Answers may vary depending on how close the relative humidity obtained is

   to 100%.

2. Applying Conclusions If you wet the back of your hand, would the water evaporate and cool your skin?

   Unless the relative humidity is 100%, at which point the net condensation

   equals evaporation, yes.

_____

EXTENSION

1. Making Inferences Suppose that you exercise in a room in which the relative humidity is 100%.

a. Would the moisture on your skin from perspiration evaporate easily?

no _____

_____

b. Would you be able to cool off readily? Explain your answer.

no; because the air is saturated no more water will evaporate from the skin

and, thus, there will be little cooling effect

_____

2. Applying Ideas Suppose that you have just stepped out of a swimming pool. The relative humidity is low, about 30%. Would you feel warm or cool? Explain your answer.

cool; because the relative humidity is low, rapid evaporation would cause the

skin to cool

_____

_____

Dew Point

Teacher's Notes

Provide lab thermometers so students can determine the current dew point of the air. Tell them to record the starting air temperature next to the glass and the temperature when a thin layer of moisture (dew) forms on the outside of the container.

SKILLS ACQUIRED

Experimenting
Observing
Analyzing
Communicating

Quick Lab

Dew Point

MATERIALS
- water (room-temperature)
- glass container
- ice cubes (one or two)

SAFETY

PROCEDURE

1. Pour **room-temperature water** into a **glass container**, such as a drinking glass, until the water level is near the top of the cup.
2. Observe the outside of the glass container, and record your observations.

_____

_____

3. Add **one or two ice cubes** to the container of water.
4. Watch the outside of the container for five minutes for any changes.

ANALYSIS

1. What happened to the outside of the container?

   Moisture appears outside the container.

   _____

2. What is the liquid on the container?

   water

   _____

3. Where did the liquid come from? Explain your answer.

   from the surrounding air; the ice lowered the temperature of the air to

   below the dewpoint, which caused the water vapor to condense on the cool

   surface of the container.

Cloud Formation

Teacher's Notes

Substitute a metal dish for the jar lid, and use additional ice cubes to chill the air above the water's surface more rapidly. To introduce the concept of condensation nuclei, use matches to add smoke particles to the jar, and then seal the jar quickly. Have students compare the size and speed of cloud formation.

SKILLS ACQUIRED
Experimenting
Observing
Analyzing

Name _____ Class _____ Date _____

Cloud Formation

MATERIALS

- bottle opener
- glass jar
- 1 mL of hot water
- ice cube

PROCEDURE

1. Use a **bottle opener** to puncture 1 or 2 holes into the lid of a **glass jar.**
2. Pour **1 mL of hot water** into the jar.
3. Place an **ice cube** over the holes in the lid of the jar. Make sure the holes are completely covered.
4. Observe the changes that occur within the jar.

ANALYSIS

1. Draw a diagram of the jar. Label the areas of the diagram where evaporation and condensation take place. Also, label areas where latent heat is released and absorbed.

Evaporation takes place near the water's surface, where latent heat is

absorbed. Condensation in the form of a cloud forms at the top of the jar,

where latent heat is released.

2. Explain why latent heat was released and absorbed in the areas you labeled on the diagram.

The conversion of liquid water to a gas requires energy to break the attrac-

tive forces between water molecules. When the process is reversed, the

energy reenters the air.

Inquiry Lab

How Big Is a Raindrop?

Teacher's Notes

TIME REQUIRED one to two 45-minute class periods

LAB RATINGS   Easy ←——1——2——3——4——→ Hard
    Teacher Prep–2
    Student Set-Up–2
    Concept Level–2
    Clean Up–2

Rebecca Grella
Brentwood
High School
Brentwood, New York

SKILLS ACQUIRED
    Collecting Data
    Communicating
    Constructing Models
    Experimenting
    Interpreting
    Organizing and Analyzing Data
    Predicting

THE SCIENTIFIC METHOD

In this lab students will

- Make Observations

- Ask Questions

- Test the Hypothesis

- Analyze the Results

- Communicate the Results

MATERIALS

It is recommended that students work in pairs or groups of three or four for this lab. Metal or plastic ice cube trays (dividers removed), aluminum baking pans or clean deli containers work well. Open plastic petri dishes may also be used.

TIPS AND TRICKS

Wait for a rainy day to perform this experiment. This lab has links to a series of experiments done in the early 1900s by a Jericho, Vermont farmer who also photographed snowflakes. His name was Wilson A. "Snowflake" Bentley. Bentley found that dough pellets would form in a pan of flour when raindrops hit the flour. If he let these pellets dry, he could then remove them from the flour and measure their size. The dough pellets produced were always very close in size to the raindrops that made them. Bentley reported the results of his raindrop research in *Monthly Weather Review* in October, 1904, in an article titled "Studies of Raindrops and Raindrop Phenomena." From the data he amassed during his experiments, he concluded that drops of all sizes were usually present in most rains. However, smaller drops usually outnumbered the large ones, except during the drenching rains of a thunderstorm. Giving students the background of this experiment after they perform this lab allows them to place their experience into a historical context.

Teachers may make laminated placemats by laminating together one black and one white sheet of construction paper. In addition to making cleanup easier, the white flour pellets may be placed on the black placemat, making for easier observations. The placemats may be stored for other experiments.

Inquiry Lab

Using Scientific Methods

How Big Is a Raindrop?

Have you ever been caught in the rain? If you have, then you know that some raindrops are smaller than others. When it's drizzling, most of the drops seem to be very small. When there is a downpour, most of the drops seem to be larger. So, how big is a raindrop? Are all the drops that fall during a particular episode of rain all the same size? In this lab, you will discover some answers by doing an experiment with real raindrops.

OBJECTIVES

Form a hypothesis about the size of raindrops.

Investigate the size of raindrops by performing an experiment.

Observe raindrops.

Collect data about real raindrops as they fall.

Draw conclusions about the size of raindrops.

MATERIALS

- beaker or cup
- dropper
- flour (approximately1 cup)
- meterstick or ruler
- pan, metal or plastic
- paper plate or laminated placemat
- sifter or strainer, flour
- smock or old shirt, one for each group member
- water

SAFETY

ASK A QUESTION

1. How big is a raindrop?

FORM A HYPOTHESIS

2. Form a hypothesis that answers your question. Explain your reasoning.

Answers may vary. Sample answer: Raindrops will vary in size even in a

single episode of rain. Drop size varies because of different sizes of conden-

sation nuclei and different rates of coalescence.

67

How Big is a Raindrop? *continued*

TEST THE HYPOTHESIS

3. Wait for a rainy day to perform this experiment.

4. Wear a smock or old shirt to protect your clothing from the flour.

5. Do all the following steps with the pan on the placemat or paper plate. Sift the flour into the pan evenly. When you are finished, the flour should be about one inch deep in the pan. The surface of the flour should be nearly level. **CAUTION:** Spilled flour may cause you to slip and fall. Wipe up any spilled flour immediately.

6. Before you do the experiment with real raindrops, do a test run with artificial raindrops. Fill the dropper with water. Squeeze the dropper slightly, so that a drop of water forms, *but does not fall*. Measure the drop as best you can with the meterstick or ruler. Record your measurement.

Answers may vary. _____

_____

_____

7. After you record your measurement, hold the dropper as high as you can above the pan. Gently squeeze the dropper until the drop falls. What happens when the drop hits the flour in the pan?

When the drop hits the pan, a small dough pellet forms. _____

_____

_____

_____

8. Observe the dough pellet. How does the dough pellet compare in size to the drop of water?

The dough pellet and the original water drop should be about equal in size. _____

_____

_____

_____

9. Repeat Steps 6, 7 and 8 two more times. What did you observe about the results when you repeated the experiment?

Answers may vary. The dough pellet and the water drops were about equal in

size. _____

_____

How Big is a Raindrop? *continued*

10. Remove the dough pellet from the pan and set it aside. Place the flour pan out in the rain for four or five seconds, then bring it inside again. What happened to the flour in the pan when it was exposed to the raindrops?

Answers may vary. Sample answer: A dough pellet formed wherever a rain-

drop hit the flour. The dough pellets were or were not all the same size.

11. Measure each of the pellets that formed and record the measurements in the space below.

Answers may vary.

ANALYZE THE RESULTS

1. Analyzing Results How did the dough pellets produced by the raindrops compare in size to each other?

Answers may vary. Answers may include that the pellets were the same size,

or varied in size.

2. Explaining Events What might explain the results you obtained for the sizes of the dough pellets?

Answers may vary. Sample answer: If the raindrops were all the same size,

the dough pellets would be approximately the same size; if the size of the

raindrops varied, the size of the dough pellets probably varied.

How Big is a Raindrop? *continued*

DRAW CONCLUSIONS

3. **Drawing Conclusions** What was the purpose of doing the practice experiment with the dropper before doing the experiment with real raindrops?

   Answers may vary. Sample answer: The dough pellet should reflect the actual

   size of the water drop. _____

   _____

   _____

4. **Making Predictions** If you were to repeat the experiment during another rain shower, would you expect the raindrop-size results to be the same? Explain.

   Answers may vary. Sample answer: Depending on the kind of rain shower, the

   raindrop sizes might be different than they were the first time. _____

   _____

   _____

EXTENSION

1. **Long-term Investigation and Communication** Repeat the experiment on several other rainy days. Keep track of the results of the sizes of the raindrops. Make a poster describing what you did and illustrate your results by creating a data chart, and making drawings of the size of raindrops.

   Answers may vary. _____

   _____

   _____

   _____

   _____

   _____

   _____

   _____

What Is the Shape of a Raindrop?

Teacher's Notes

TIME REQUIRED one to two 45-minute class periods

LAB RATINGS Easy ←—— 1   2   3   4 ——→ Hard

Teacher Prep–2
Student Set-Up–2
Concept Level–2
Clean Up–1

Linda Prince
Syosset
High School
Syosset, New York

SKILLS ACQUIRED

Communicating
Constructing Models
Experimenting
Interpreting
Predicting

THE SCIENTIFIC METHOD

In this lab, students will

- Make Observations

- Ask Questions

- Test the Hypothesis

- Analyze the Results

- Communicate the Results

MATERIALS

Use only Ground Fault Circuit Interruptor protected outlets for the electrical appliances.

Make sure that no student drops water directly into any of the electrical appliances; make sure that students understand the dangers of using water near an electrical appliance. Using paper toweling or cheesecloth on top of the opening of the plastic pipe acts as a baffle to disperse the air stream. It also absorbs the drops. The air stream will dry the drops that land on the absorbent material. Another possible source of blowing air would be battery-powered, small personal fans, without plastic piping. If a battery-operated fan is used, place the fan underneath a ringstand ring that has one layer of cheesecloth draped over it. If you are performing the lab as a demonstration only, you may wish to use an air compressor.

TIPS AND TRICKS

This lab experiment may be performed as a demonstration.

Name _____ Class _____ Date _____

What Is the Shape of a Raindrop?

What shape do you think of when you picture a raindrop? Most people picture a teardrop shape, wide on the bottom and pointy on the top. Is that the actual shape of a raindrop? In this experiment, you will create model raindrops and observe their shapes. You will find out the true shape of a raindrop.

OBJECTIVES

Using Scientific Methods **Model** the way raindrops behave in a cloud.

Investigate the shape of raindrops by performing an experiment.

Using Scientific Methods **Observe** the shape, or shapes, of raindrops.

MATERIALS

- beaker or cup
- cheesecloth or paper towel
- clamp
- dropper
- plastic flexible pipe
- ring stand
- rubber band
- vacuum cleaner or hair dryer
- water
- wide clear tape or duct tape

SAFETY

PROCEDURE

1. Attach the plastic pipe to the exhaust port of the vacuum cleaner, or the hair dryer set on "cold." Attach the open end of the pipe to the ring stand, using the clamp. Place a layer of cheesecloth or paper towel over the opening of the plastic pipe. Secure the cheesecloth or paper towel with the rubber band. Tape the vacuum tube to the base of the ring stand or countertop to keep the force of the air from flipping the ring stand.

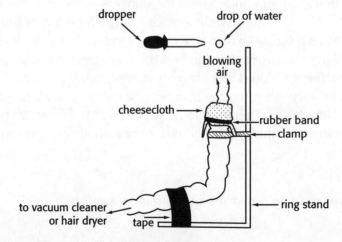

What is the Shape of a Raindrop? *continued*

2. Turn on the vacuum cleaner or hair dryer.

3. CAUTION: Never place drops of water directly over an electrical appliance. There is a danger of electrocution.

Fill the dropper with water. Squeeze one drop of water into the air stream. The drop should be suspended in the air stream, just as a drop of water in a cloud would be suspended by upward air currents. If the air stream is too strong to suspend a drop, place another layer of cheesecloth or paper toweling over the opening of the plastic pipe.

CAUTION: Use only water droplets in the wind stream as directed —do not use any other materials. Wipe up any spilled water immediately.

4. Observe the shape of the "raindrop." Write your observations below.

Answers may vary. Sample answer: The raindrops were spheres or flattened

spheres.

_____

_____

_____

_____

5. Draw a small illustration to show the shape of the raindrop.

What is the Shape of a Raindrop? *continued*

6. Repeat the experiment two more times. Report your observations in the space below.

Answers may vary. Sample answer: The raindrops were spheres or flattened

spheres.

_____

_____

_____

ANALYSIS AND CONCLUSION

1. Explaining Events Describe what happens when you place a drop of water into the air stream.

Answers may vary. Sample answer: The drop becomes suspended and looks

like a sphere or flattened sphere.

_____

_____

2. Analyzing Results Did the results of the experiment match the shape that you thought it would be? Explain.

Answers may vary.

_____

_____

_____

3. Evaluating Models How do you think this model compares to the natural processes that determine the shape of a raindrop? Explain.

Answers may vary. Sample answer: This model seems similar to the natu-

ral processes that shape a drop because in both cases air pushing up might

cause a falling drop to flatten.

_____

_____

What is the Shape of a Raindrop? *continued*

EXTENSION

1. Making Models Repeat the experiment using different-sized drops of water, and/or different air speeds. Report the results to your class.

Answers may vary.

_____

_____

_____

_____

2. Research and Communication Many scientists are interested in studying the shape of raindrops. NASA has even done experiments about the shape of raindrops in microgravity on the International Space Station. Do research about the shape of raindrops on the internet or in the library. Report your results to the class.

Answers may vary.

_____

_____

_____

_____

Internet Activity

CLIMATE AND PRECIPITATION

1. Answers may vary.
2. Answers may vary. Sample answers include the name of a region + climate, such as "Pacific Northwest climate" or "Minneapolis climate."
3. Answers may vary. Answers should indicate the correct URLs for Web sites that describe the climate of a region.
4. Answers may vary. Answers should correctly identify the sources of the information on the sites.
5. Answers may vary. Answers should include any universities and professional organizations affiliated with the Web sites.
6. Answers may vary.
7. Answers may vary.
8. Answers may vary.
9. Answers may vary.

LATENT HEAT AND THUNDER-STORMS

1. Answers may vary.
2. Answers may vary. Sample answer: latent heat thunderstorms
3. Answers may vary. Answers should indicate the URLs for Web sites that contain data about latent heat.
4. Answers may vary.
5. Answers may vary.
6. The latent heat from the condensation of warm humid air keeps the air inside a cloud warmer than the air around it. The cloud continues to develop in updrafts as long as heat energy fuels it.
7. The release of latent heat warms the rising air, causing differences in air density. Different wind speeds at different levels of the atmosphere cause the air rise at extreme speeds, leading to the winds of a tornado.
8. Diagrams may vary, but should show different stages of a storm and its motion.